The Unofficial Robotic Raspberry Pi 4B Picture Guide Part 1

By Michael Sebren

Table of Contents

Preloading the Raspberry Pi Operating System   7

Finally We're Ready to Start-Up the Pi!   23

Setting Up the Pi   36

Necessary "Part" List.

- 1 Windows PC
- 1 Raspberry Pi 4B 4Gigabytes with kit.
  - 1   RPi
  - 1   Power Supply
  - 1   HDMI micro to normal
- 1 WIFI Router.
- 1 Internet Connection.
- 1 Battery Power Supply
- 1 microSD card.
- 1 MicroSD card reader.
- 1 Television set with HDMI input.
- 1 USB Keyboard
- and 1 USB Mouse.

For my Raspberry Pi Robot, all of the above parts were included. They are the bare minimum required.

Connect WIFI Router to Internet

This one is fairly straightforward. I just plugged a Cat 5 cable from the router's internet connection, into the modem's output. And the router connected to the internet automatically. The hard part was setting setting the WIFI password.

You will need the WIFI password later for connecting to the Raspberry Pi.

You can write the WIFI login below if you need to:

WIFI Name: _____
Password: _____

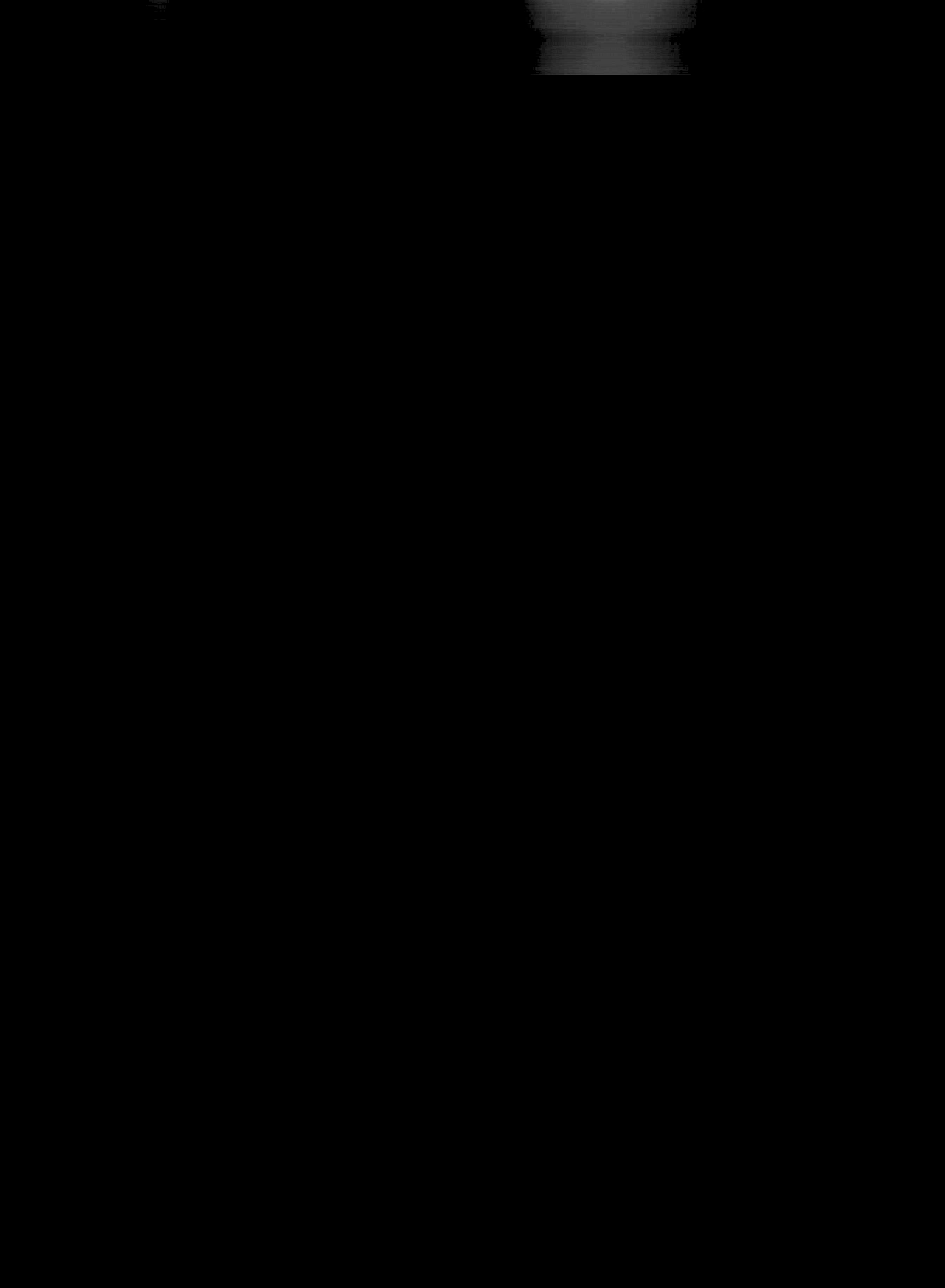

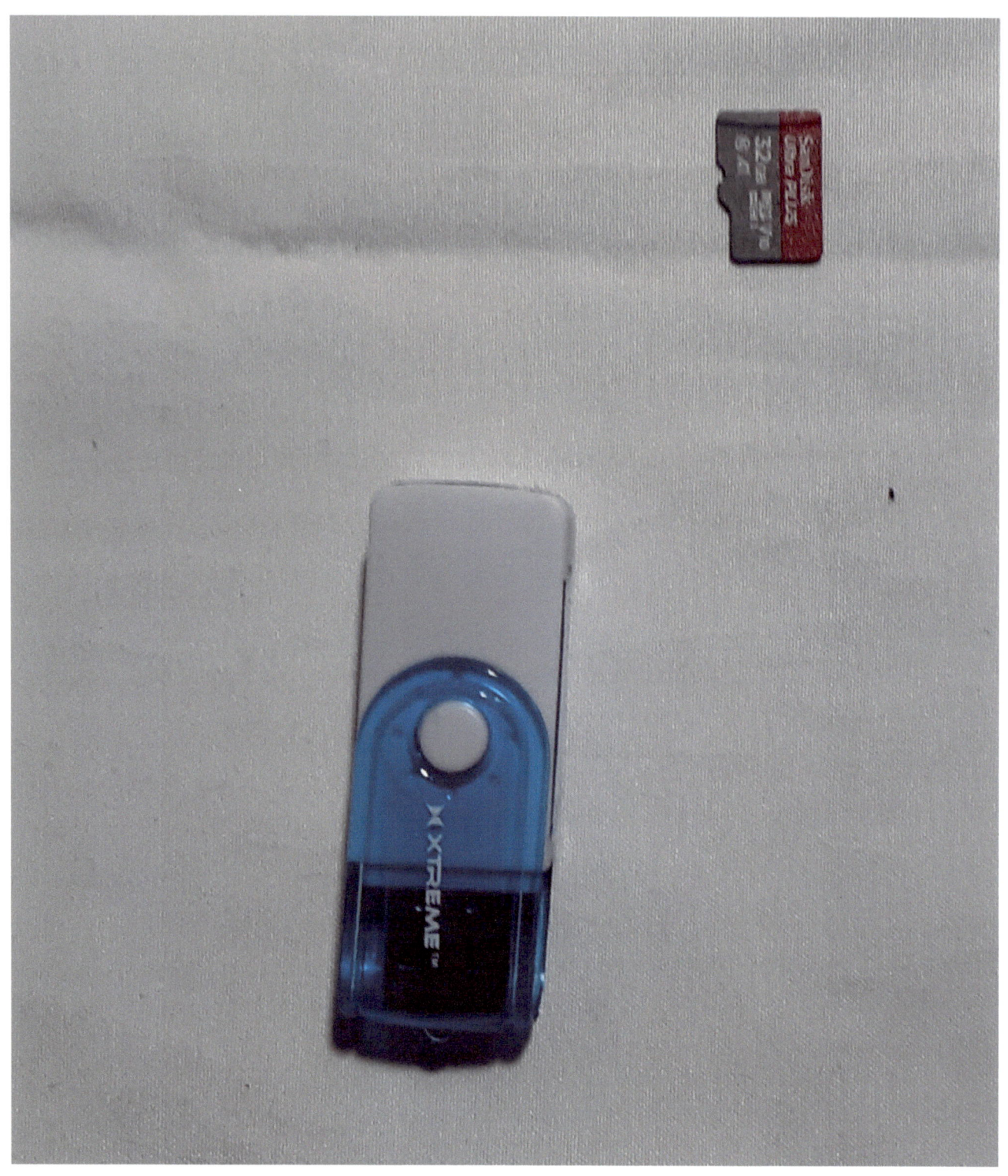

MicroSD Card Reader (USB)

The microSD card goes into the card reader; and that goes into the Windows PC.

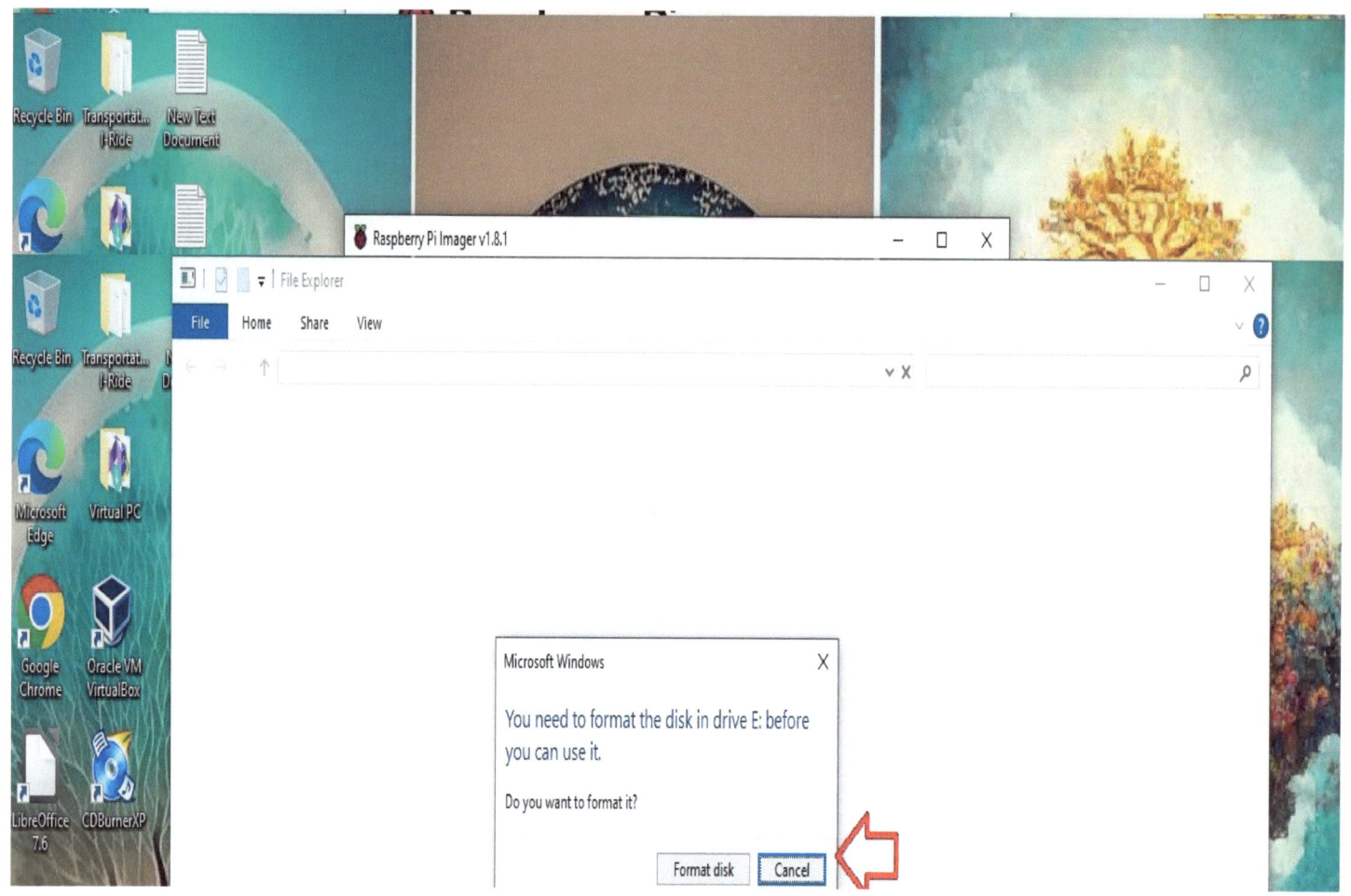

Do Not Format Using Windows

Windows will try to format automatically. Cancel those format requests.

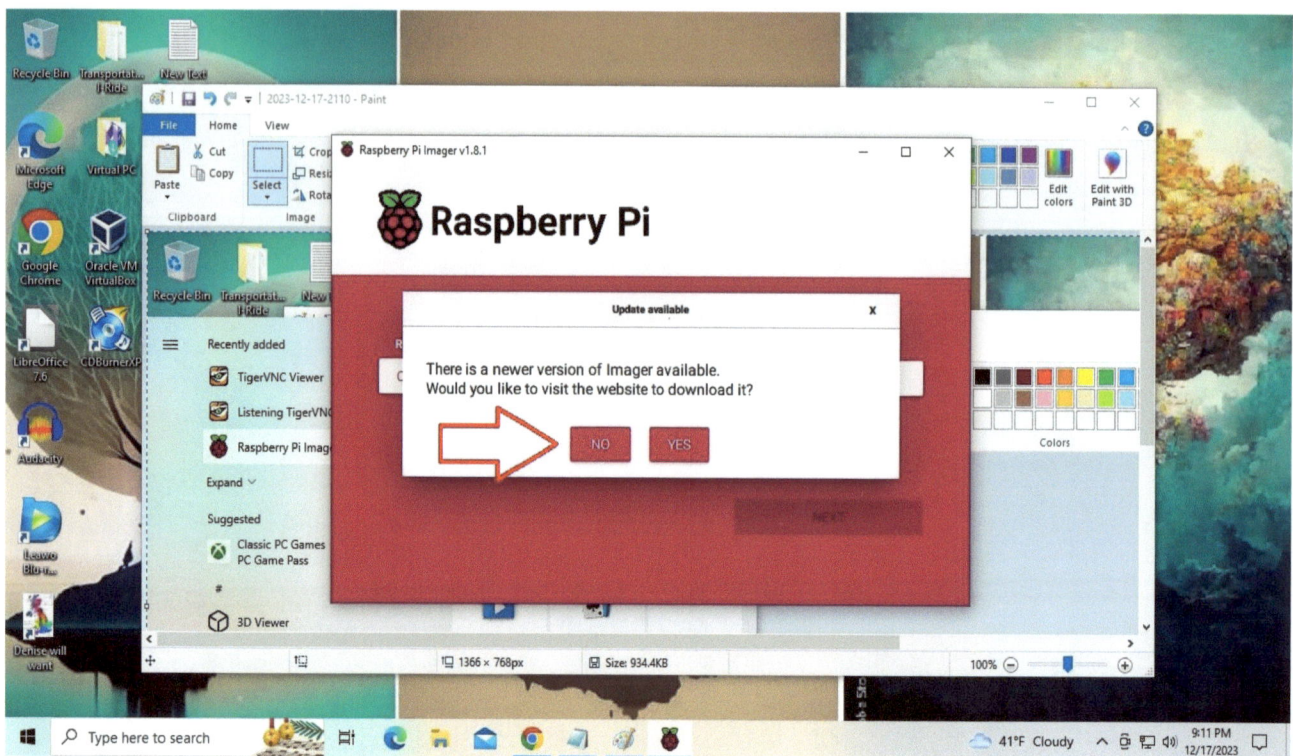

You Don't Need to Update Imager

This page probably won't come up, because a fresh install of the imager will be up to date.

If it does ask, click, "No".

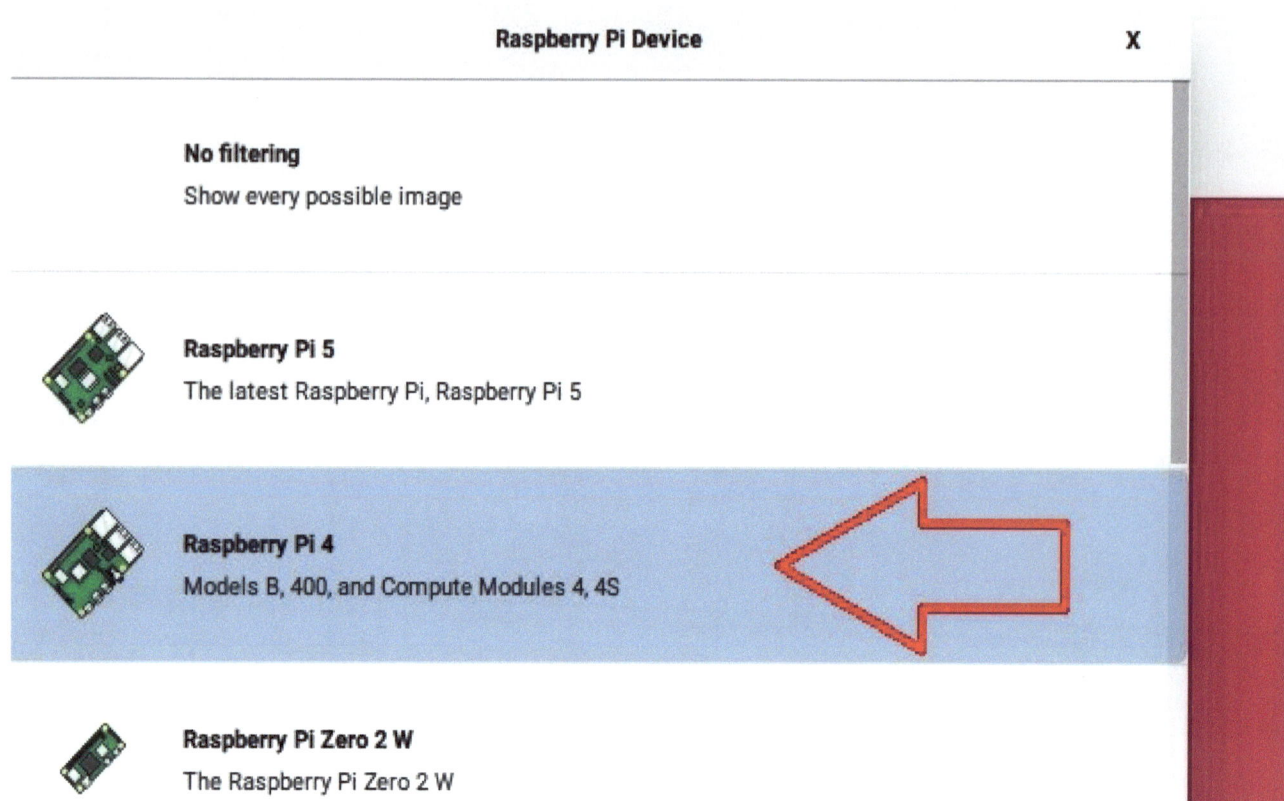

Click On Raspberry Pi 4

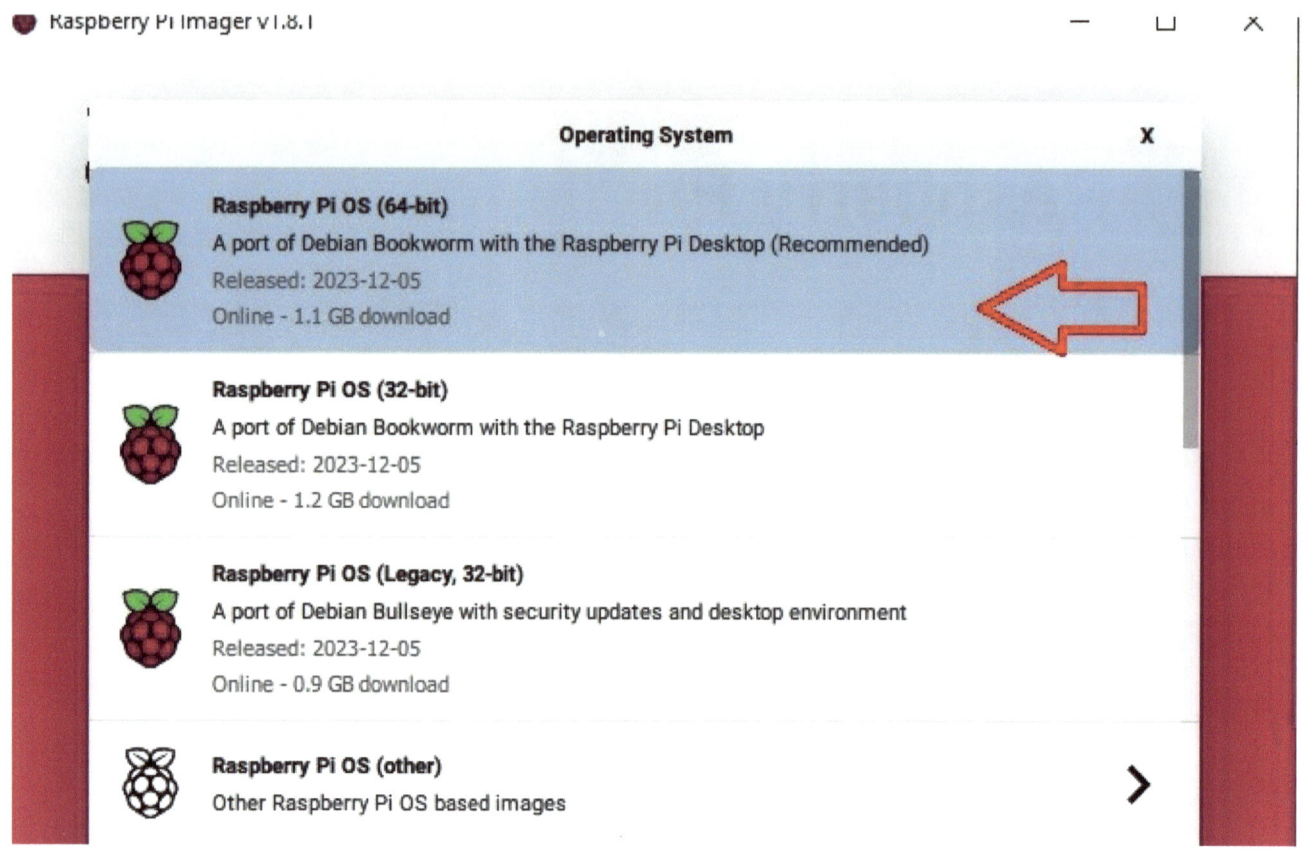

Click On Raspberry Pi OS (64-bit)

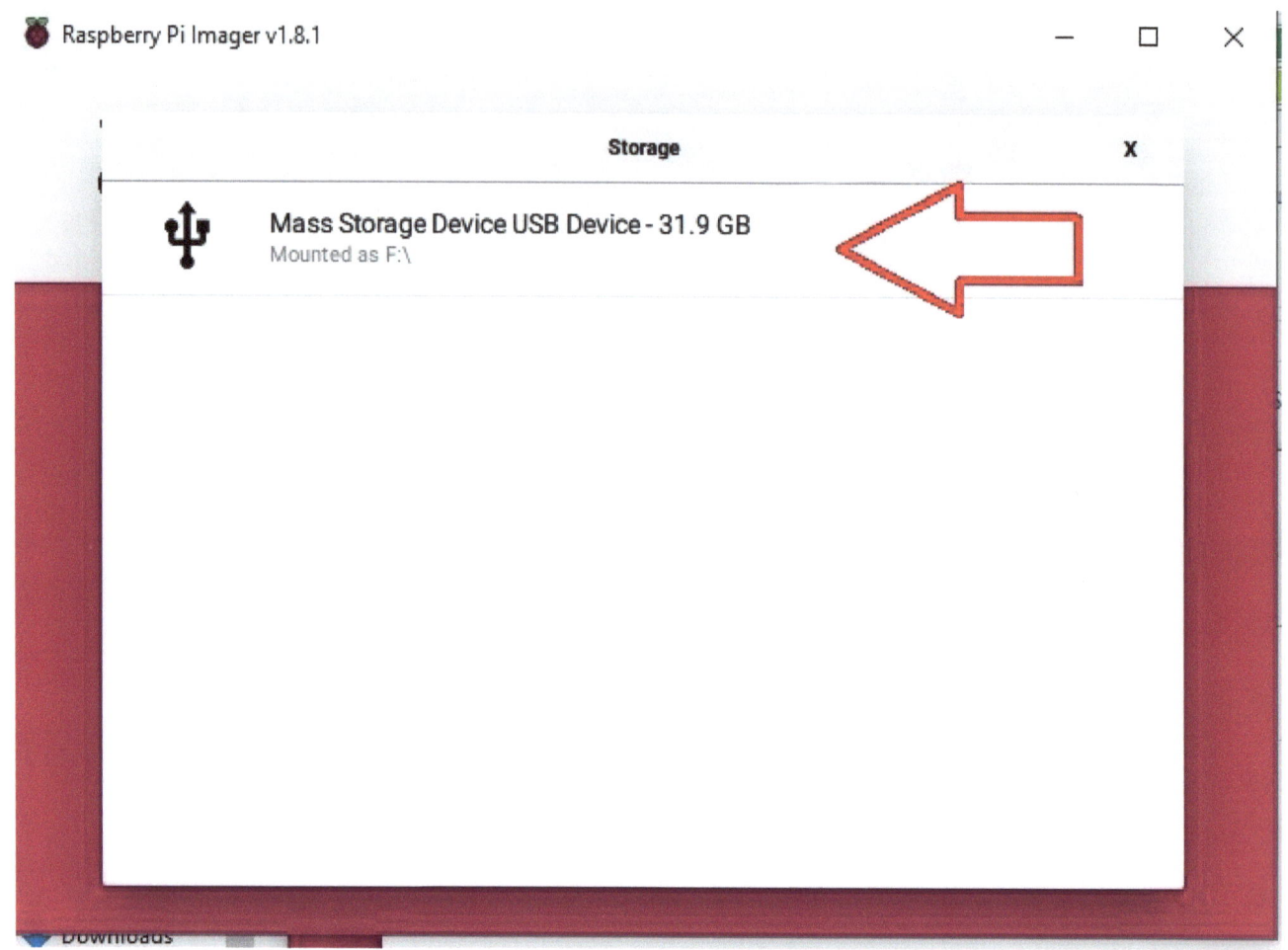

Hopefully There's Only One Storage Device

If you know the name of your card reader, then that might help.

Otherwise, it's going to be based on the gigabytes of storage on the MicroSD card.

Mine was 32GB. 31.9 is close enough. What is your device capacity?

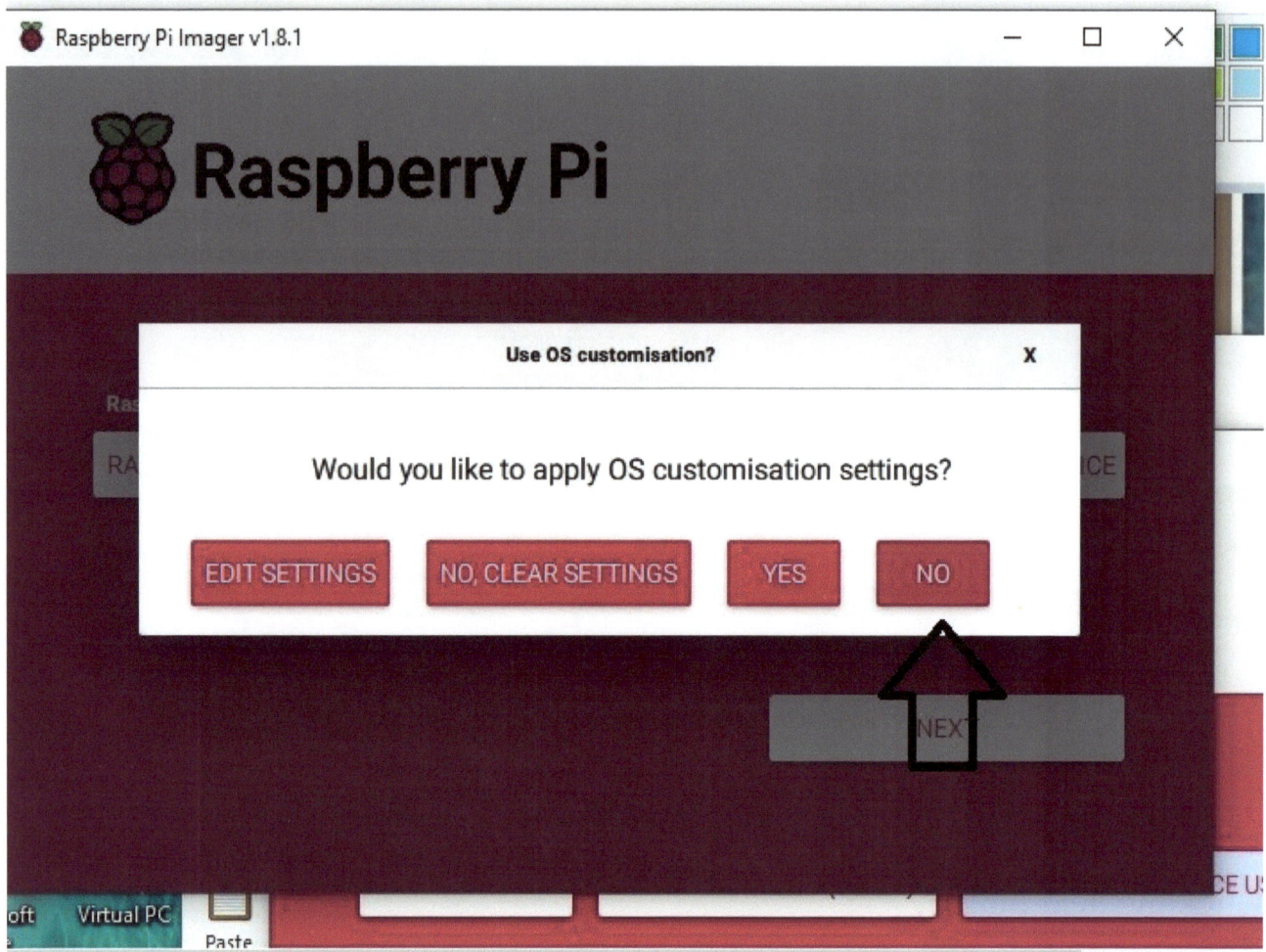

You Probably Want No

More advanced users may want to edit settings now, but most people can do it on the Raspberry Pi itself.

I suggest doing the settings after the microSD card is finished recording.

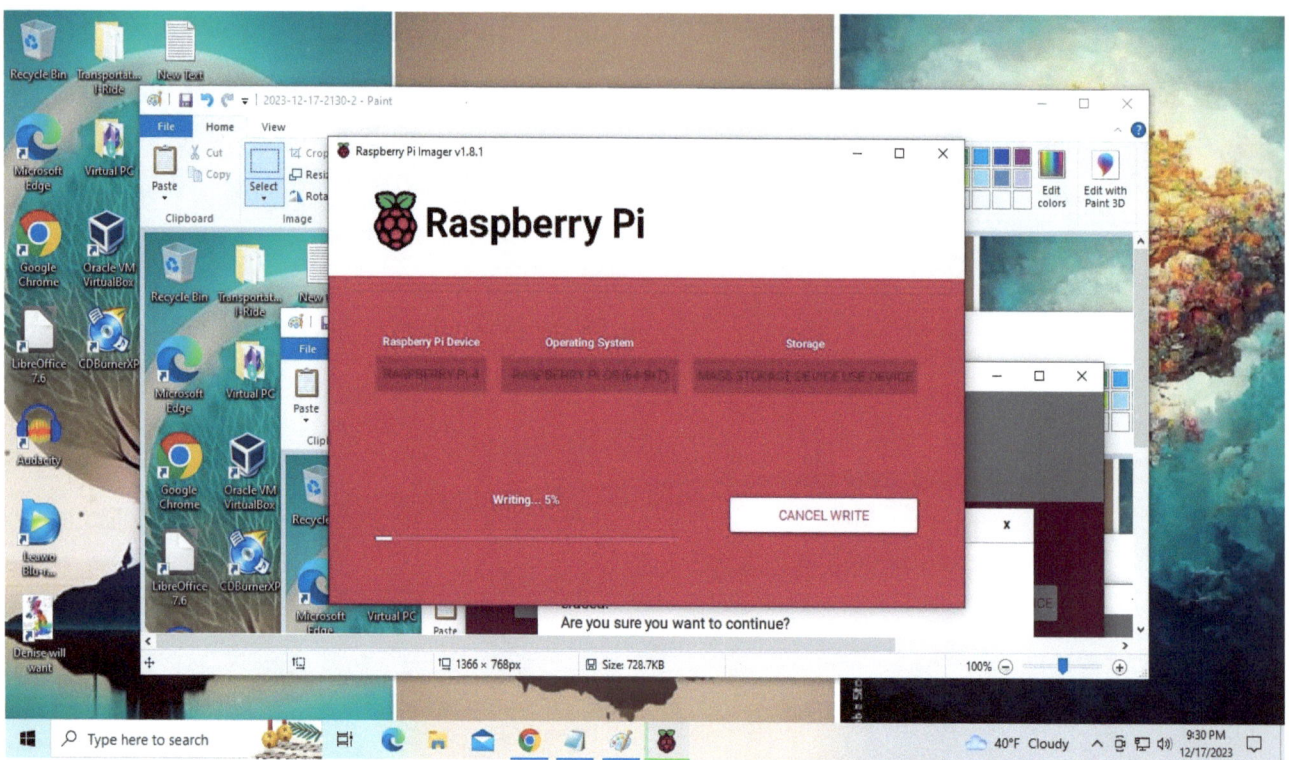

Wait…And Wait Some More…

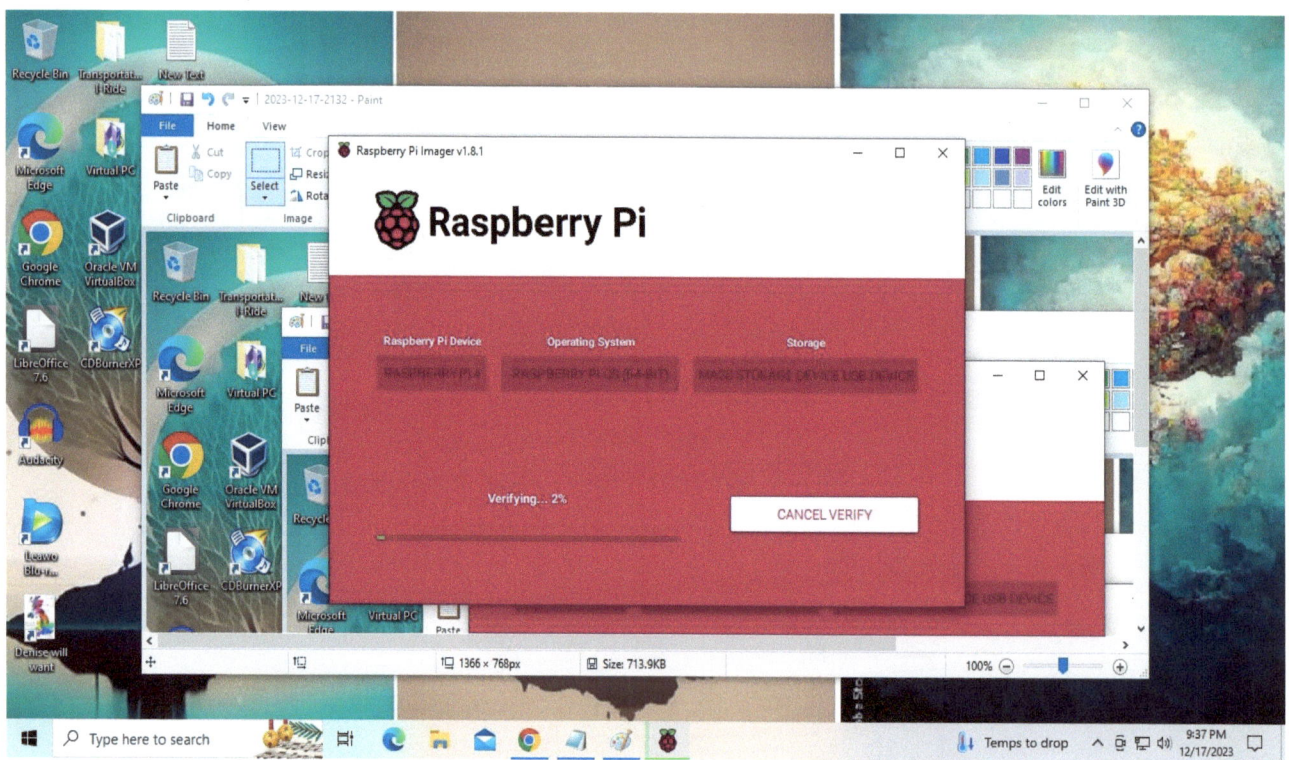

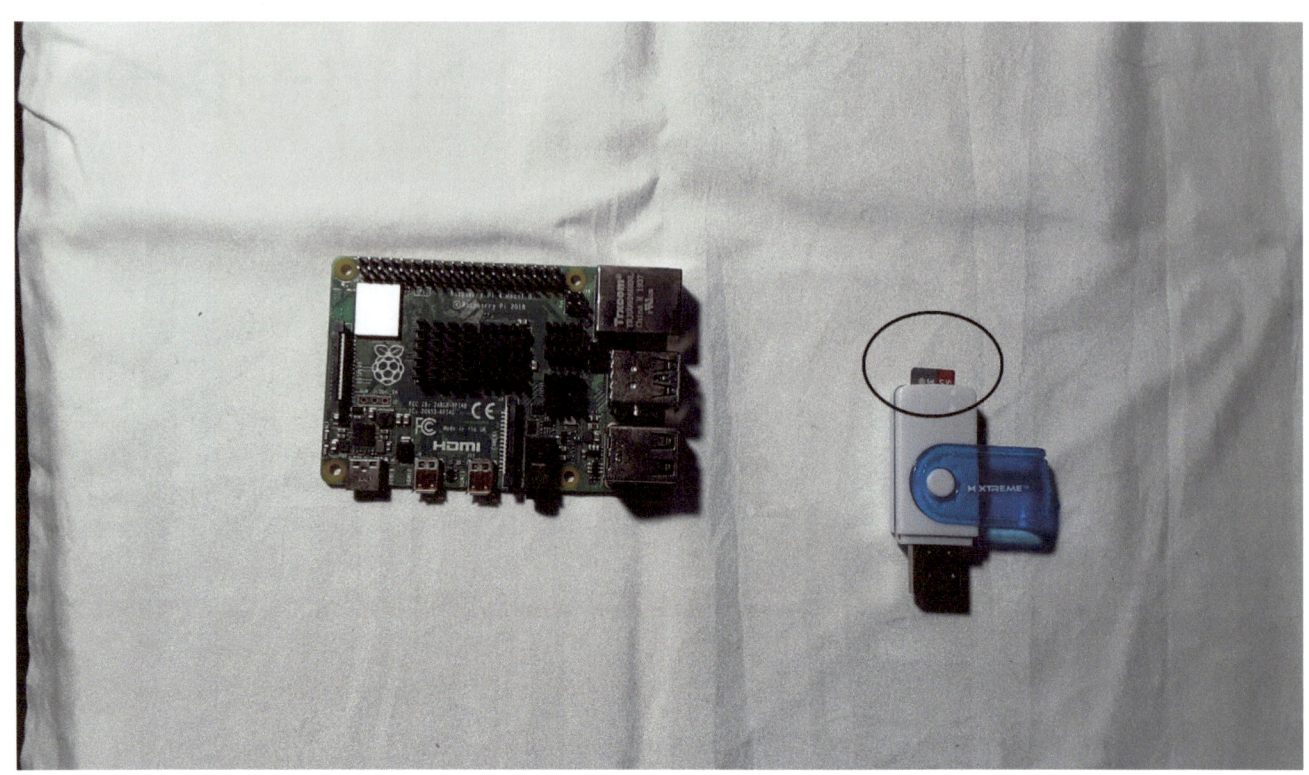

After You Have Removed the MicroSD Card Reader

The card will be in the reader, as circled.

You need to remove the microSD card and set it down.

Turn It Over; Put the Card In

This is a very delicate process, so be gentle; and, put the card into the bottom of the Raspberry Pi.

Let's Attach an HDMI Cable

Micro HDMI Attached

Be careful with the HDMI port. It's delicate.

And it will only go in one way, as shown.

Power Supply On the Left

This is what it should look like with the power supply and the HDMI connected.

You may now plug the other side of the HDMI in.

Then plug in the converter to the wall.

There They Are Both Attached

You won't see the other one in the picture, because it is beneath.

Setting Up the Pi

When you first turn on the Raspberry Pi 4B, it should boot using the operating system which was installed onto the microSD card.

Like many operating systems, it requires some information and setup questions to be answered. The following are an example of those preliminaries.

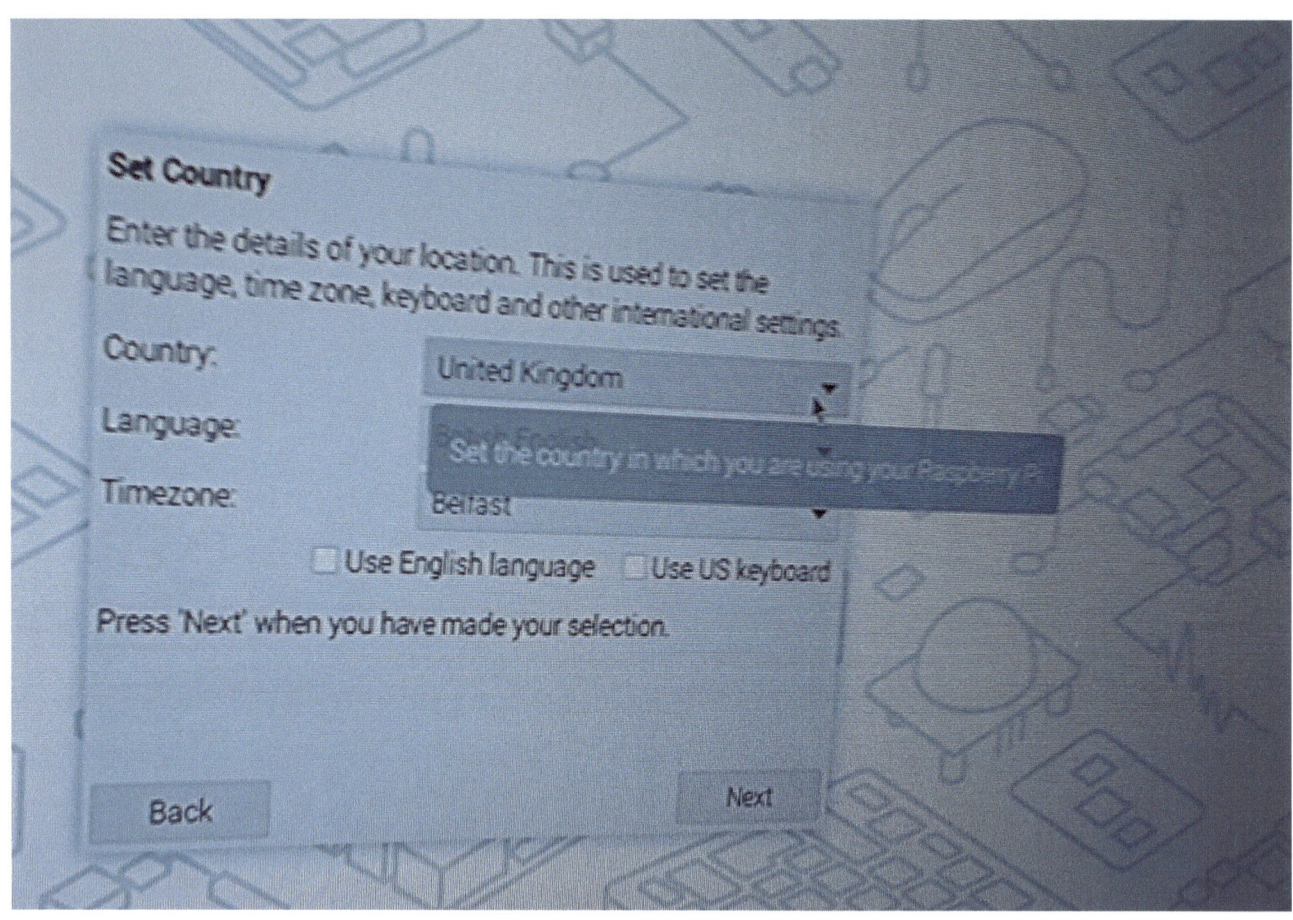

Setup Country Settings

Change country to United States.

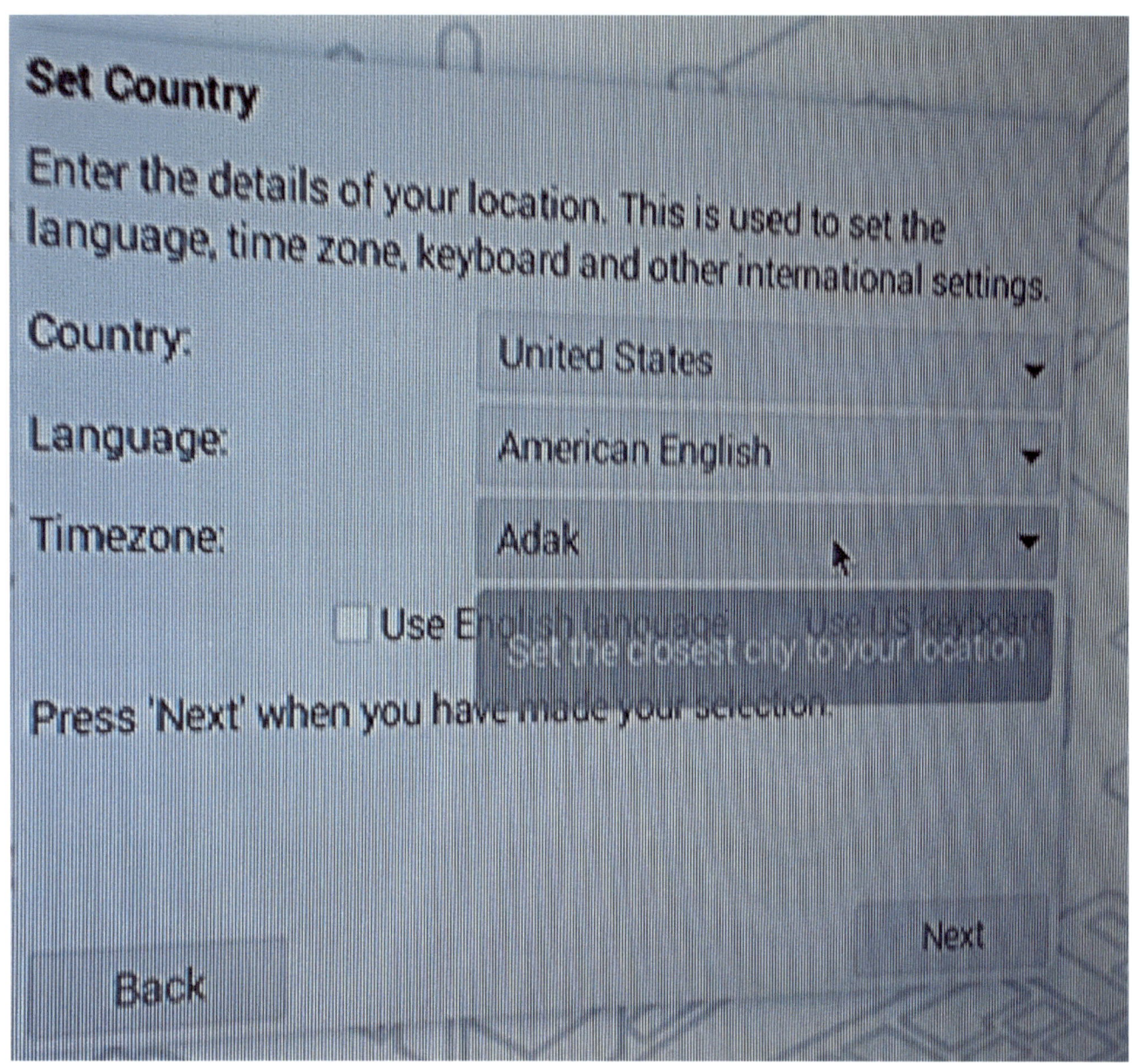

What's adak, Precious?

You don't have to keep adak time. Just click on it and choose something closer to your time zone.

For example, I live close to Detroit; or, you know, 20 miles away.

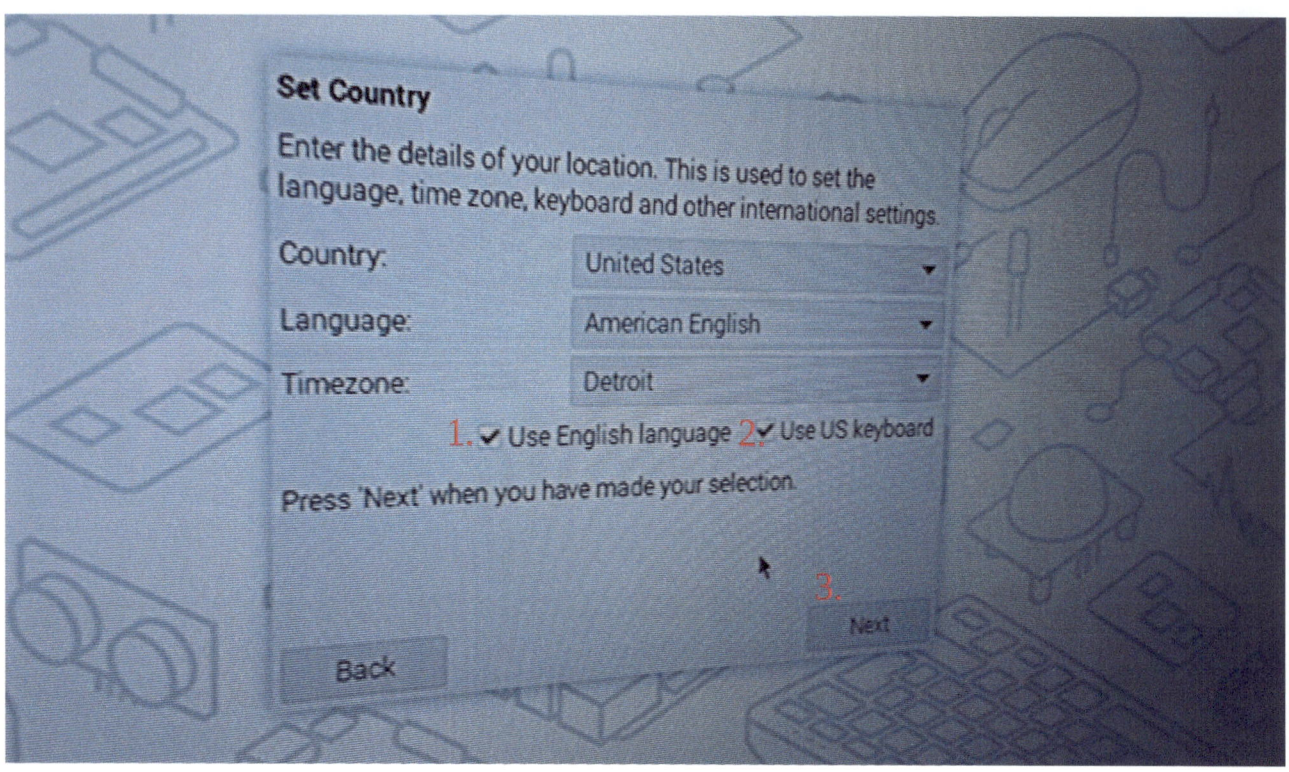

So Easy As, 1-2-3.

1. Check mark "Use English language".
2. Check mark "Use US keyboard".
3. Click "Next".

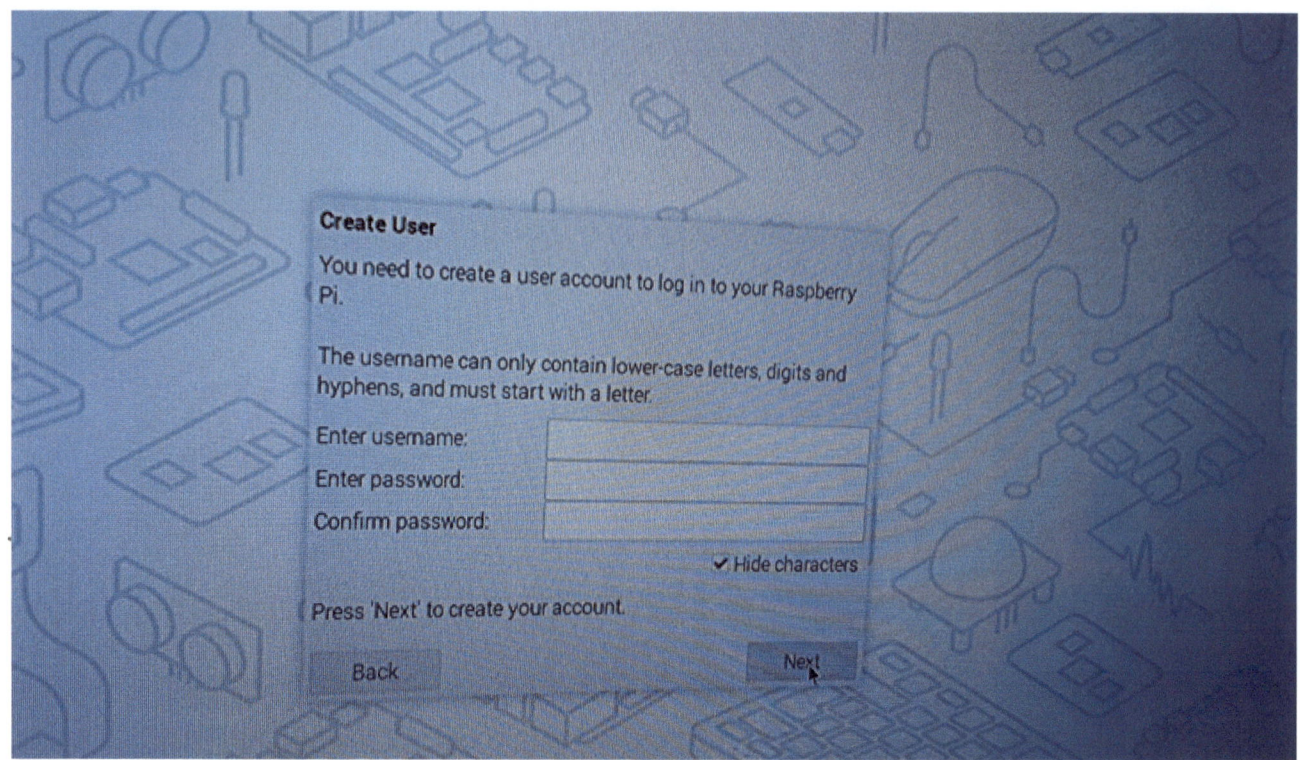

You're Gonna Want to Write These Down

Here's some space…

Username: _____

Password: _____

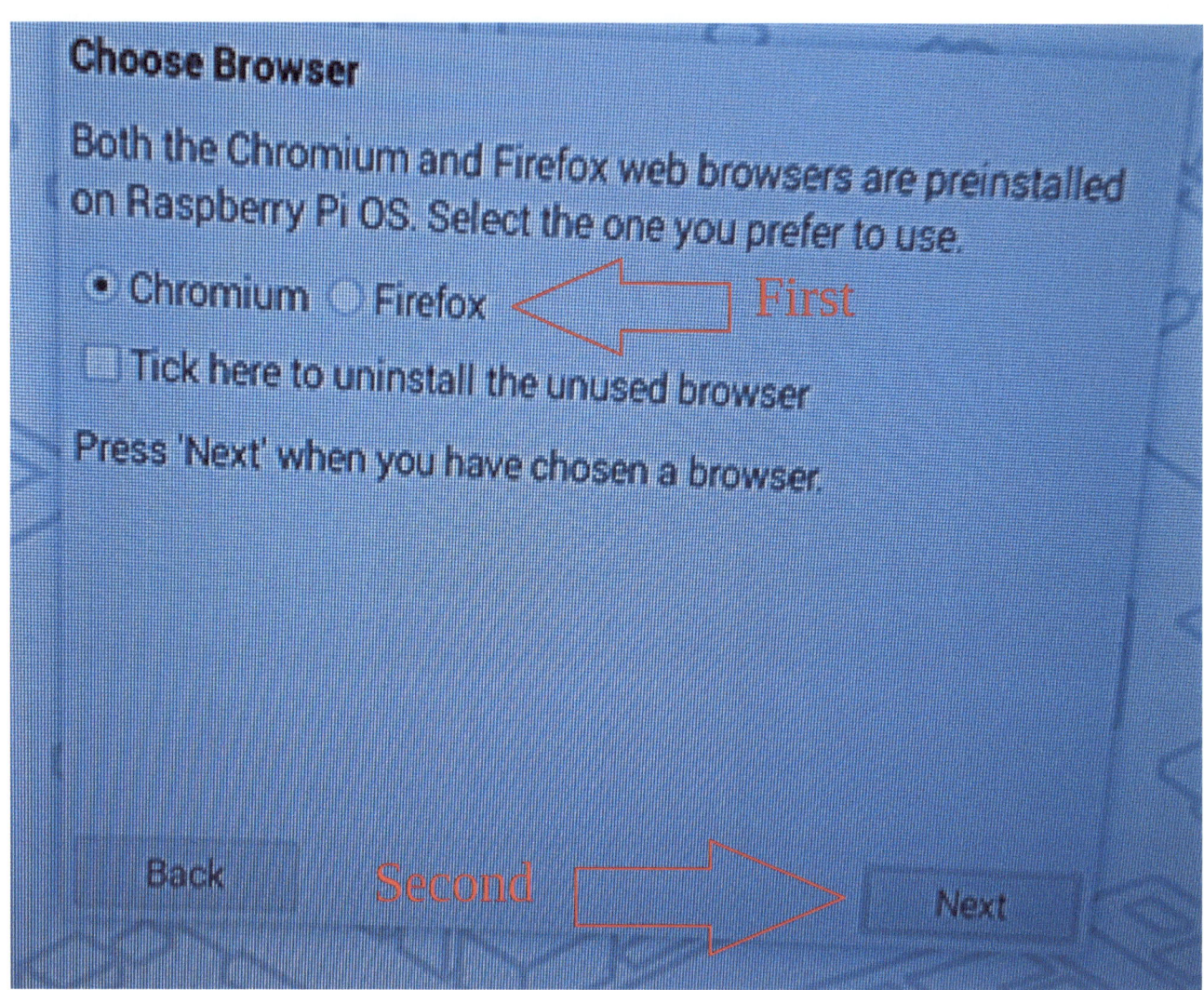

Choose Firefox

I mean, you could choose Chromium, but that's not what I'm suggesting.

Then click "Next".

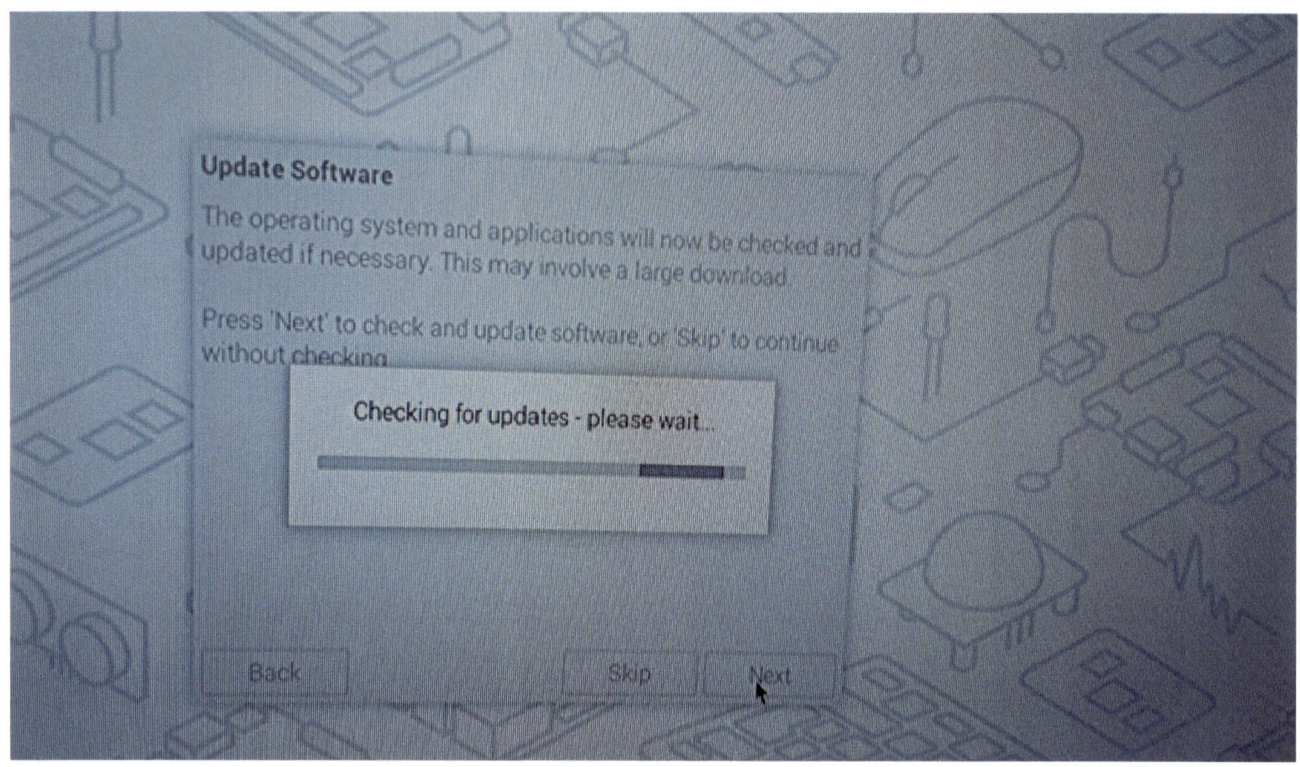

Please Wait, No Kidding

I'm sorry; but, this is going to take a while.

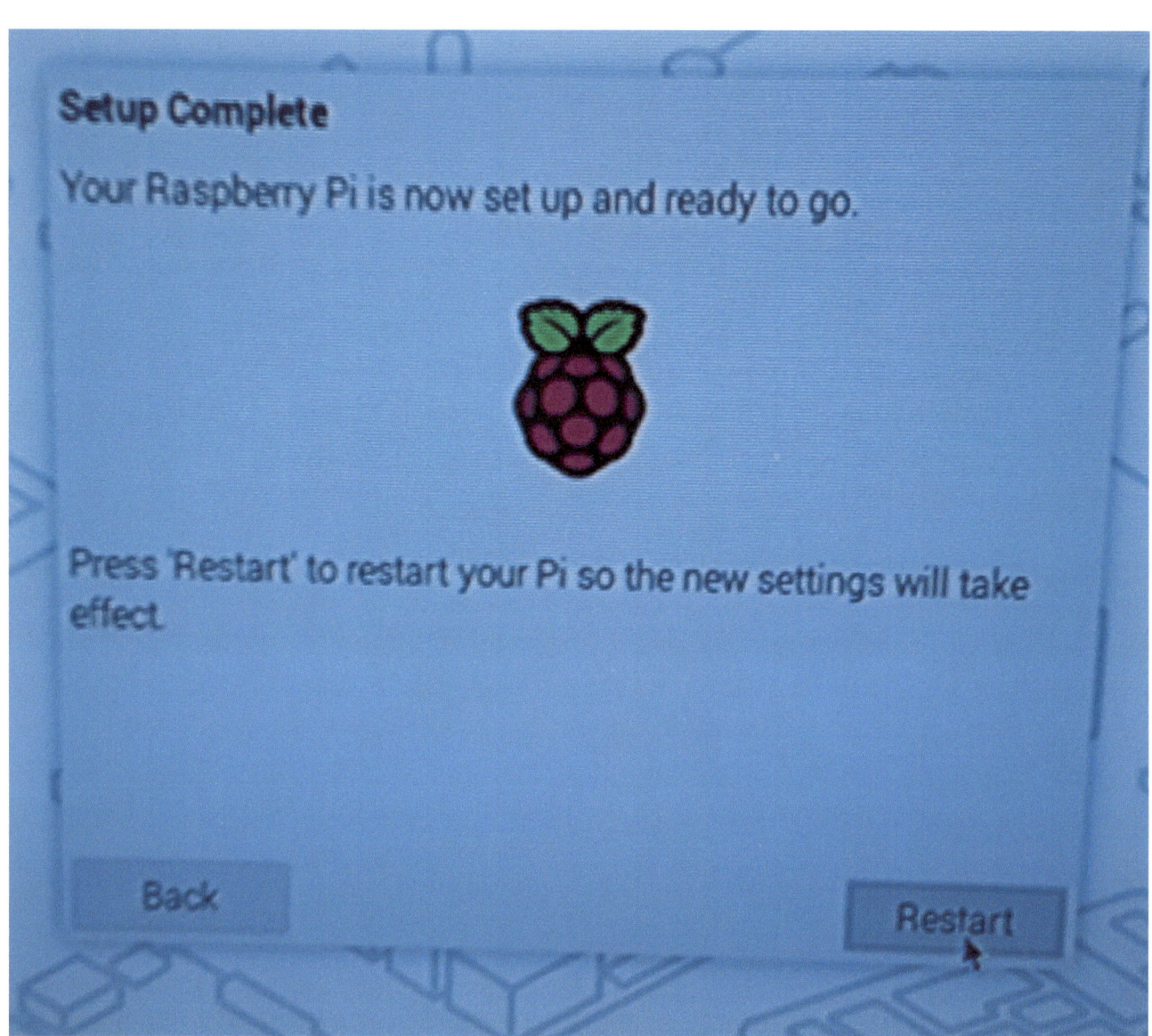

Setup Complete

Now just press the "Restart" button, and wait for a few minutes for the Pi to reboot.